Ray's New Primary Arithmetic Workbook

Book Two

Rudolph Moore, Ph.D.
Betty Moore, M.A.

© **Mott Media, Inc., Publishers**
Milford, Michigan

Procedures for this Classic Curriculum Workbook:

Classic Curriculum is designed with the student in mind. It requires the use of other Classic Curriculum materials. Follow these steps:

1. Teacher   *Remove* the test and answers from the middle of the workbook.

2. Student   *Read* Table of Contents.

                         Read "I can" sections of each lesson to get idea of material to be studied.

                         Write new or unusual words with their definitions in a notebook.

                         Study all vocabulary words.

                         Complete the workbook lesson by lesson beginning on page 2.

                         Score the work each day using only a RED pen, marking only the incorrect problem's number.

                         Rework and rescore all missed problems.

                         Devote one day to each Review and Quiz section.

                         Complete each quiz like a test. Review items missed.

                         Give the workbook to the teacher when it has been completed and test the following day.

                         Take the final test.

3. Teacher   *Score* the test. A score of at least 80% should be made before starting with the next workbook.

Mott Media Textbooks needed to complete this workbook:

☑ RAY'S NEW PRIMARY ARITHMETIC

Table of Contents

Ray's Arithmetic Workbook

Removing Objects

I can remove objects from a total.

READ

We have learned to add objects. When we add, we combine objects. We can take two or more small groups and put them (combine them) all in one big group.

Another thing we can do in arithmetic is to start with the big group. Then we can remove some objects **from** the big group. We then ask how many we have **left**.

STUDY

Big group   Remove 2   1 left

Read the activities on page 27, Lesson XXIV in your arithmetic textbook, *Ray's New Primary Arithmetic.*

RECITE
Write how many are left.

1. How many birds are left? _____

2. Remove 4 from 7. How many left? _____

3. Remove 3 from 9. How many left? _____

Complete activities 1 through 6 in Lesson XXIV, page 27 using objects. Your teacher/tutor should read the questions to **you.**

Subtraction

I can subtract using the number one.

READ

In Lesson XXV, we are learning how to subtract, or remove, **one from** a first number. Some of the sentences are removing, or subtracting, some number so that the answer is **one**.

Practice saying each of these sentences at the top of page 28, using objects. You need to learn each of these sentences so go over them many times.

STUDY

1 from 3 leaves 2   or   2 from 3 leaves 1

RECITE
Write the numbers on the lines.

1. 1 from 1 leaves _____

2. 1 from 2 leaves _____

3. 1 from 3 leaves _____

4. 1 from 4 leaves _____

5. 1 from 5 leaves _____

6. 1 from 6 leaves _____

7. 1 from 7 leaves _____

8. 1 from 8 leaves _____

9. 1 from 9 leaves _____

10. 1 from 10 leaves _____

11. 0 from 1 leaves _____

12. 1 from 2 leaves _____

13. 2 from 3 leaves _____

14. 3 from 4 leaves _____

15. 4 from 5 leaves _____

16. 5 from 6 leaves _____

17. 6 from 7 leaves _____

18. 7 from 8 leaves _____

19. 8 from 9 leaves _____

20. 9 from 10 leaves _____

Answer the questions 1 through 9 in Lesson XXV, page 28 in your textbook. Ask your teacher/tutor to ask **you** the questions.

Subtraction

I can subtract using the number two.

READ

In Lesson XXVI, we are learning how to subtract, or remove, **two from** a first number. Some of the sentences are removing, or subtracting, some number so that the answer is **two**.

Practice saying each of these sentences at the top of page 29, using objects. You need to learn each of these sentences so go over them many times.

STUDY

2 from 5 leaves 3   3 from 5 leaves 2

RECITE
Write the numbers on the lines.

1. 2 from 2 leaves _____	**11.** 0 from 2 leaves _____
2. 2 from 3 leaves _____	**12.** 1 from 3 leaves _____
3. 2 from 4 leaves _____	**13.** 2 from 4 leaves _____
4. 2 from 5 leaves _____	**14.** 3 from 5 leaves _____
5. 2 from 6 leaves _____	**15.** 4 from 6 leaves _____
6. 2 from 7 leaves _____	**16.** 5 from 7 leaves _____
7. 2 from 8 leaves _____	**17.** 6 from 8 leaves _____
8. 2 from 9 leaves _____	**18.** 7 from 9 leaves _____
9. 2 from 10 leaves _____	**19.** 8 from 10 leaves _____
10. 2 from 11 leaves _____	**20.** 9 from 11 leaves _____

Answer the questions 1 through 9 in Lesson XXVI, page 29 in your textbook. Ask your teacher/tutor to ask you the questions.

Subtraction

I can subtract using the number three.

READ

In Lesson XXVII, we are learning how to subtract, or remove, **three from** a first number. Some of the sentences are removing, or subtracting, some number so that the answer is **three**.

Practice saying each of these sentences at the top of page 30, using objects. You need to learn each of these sentences so go over them many times.

STUDY

3 from 7 leaves 4   4 from 7 leaves 3

RECITE

Write the numbers on the lines.

1. 3 from 3 leaves _____   **11.** 0 from 3 leaves _____

2. 3 from 4 leaves _____   **12.** 1 from 4 leaves _____

3. 3 from 5 leaves _____   **13.** 2 from 5 leaves _____

4. 3 from 6 leaves _____   **14.** 3 from 6 leaves _____

5. 3 from 7 leaves _____   **15.** 4 from 7 leaves _____

6. 3 from 8 leaves _____   **16.** 5 from 8 leaves _____

7. 3 from 9 leaves _____   **17.** 6 from 9 leaves _____

8. 3 from 10 leaves _____   **18.** 7 from 10 leaves _____

9. 3 from 11 leaves _____   **19.** 8 from 11 leaves _____

10. 3 from 12 leaves _____   **20.** 9 from 12 leaves _____

Answer the questions 1 through 9 in Lesson XXVII, page 30 in your textbook. Ask your teacher/tutor to ask **you** the questions.

Review I

When we add, we combine objects from two or more groups into one group. We use the word **and** to show addition.

When we subtract, we remove some objects from the big group. We then ask how many we have left. The words **from** and **leaves** are used in subtraction sentences.

When we subtract **one from** a number, we remove **one from** the big number. This **leaves one less** in the group, the answer.

When we subtract **two from** a number, we remove **two from** the big number. This **leaves two less** in the group, the answer.

When we subtract **three from** a number, we remove **three from** the big number. This **leaves three less** in the group, the answer.

Write the numbers on the lines.

1. 1 from 1 leaves _____

2. 1 from 2 leaves _____

3. 1 from 3 leaves _____

4. 1 from 4 leaves _____

5. 1 from 5 leaves _____

6. 1 from 6 leaves _____

7. 1 from 7 leaves _____

8. 1 from 8 leaves _____

9. 1 from 9 leaves _____

10. 1 from 10 leaves _____

11. 2 from 2 leaves _____

12. 2 from 3 leaves _____

13. 2 from 4 leaves _____

14. 2 from 5 leaves _____

15. 2 from 6 leaves _____

16. 2 from 7 leaves _____

17. 2 from 8 leaves _____

18. 2 from 9 leaves _____

19. 2 from 10 leaves _____

20. 2 from 11 leaves _____

21. 3 from 3 leaves _____

22. 3 from 4 leaves _____

23. 3 from 5 leaves _____

24. 3 from 6 leaves _____

25. 3 from 7 leaves _____

26. 3 from 8 leaves _____

27. 3 from 9 leaves _____

28. 3 from 10 leaves _____

29. 3 from 11 leaves _____

30. 3 from 12 leaves _____

Quiz I

Write the numbers on the lines.

1. 1 from   8 leaves _____   **11.** 3 from   7 leaves _____

2. 3 from   5 leaves _____   **12.** 1 from   5 leaves _____

3. 2 from   9 leaves _____   **13.** 2 from   3 leaves _____

4. 1 from   3 leaves _____   **14.** 3 from 11 leaves _____

5. 3 from   6 leaves _____   **15.** 2 from   8 leaves _____

6. 2 from   4 leaves _____   **16.** 3 from   3 leaves _____

7. 3 from 10 leaves _____   **17.** 3 from 12 leaves _____

8. 1 from   6 leaves _____   **18.** 2 from   6 leaves _____

9. 3 from   4 leaves _____   **19.** 3 from   9 leaves _____

10. 2 from 10 leaves _____   **20.** 2 from   4 leaves _____

21. Nine apples taken from ten apples leaves _____ apple.

22. Two of five birds flew away. _____ birds were left.

23. Bill spent two cents of his four cents. He has _____ cents left.

24. Mary ate three of four oranges. She had _____ orange left.

25. Eight taken from eleven leaves _____ .

Answer five oral questions from Lessons XXV, XXVI, or XXVII.
These questions should be asked by a teacher/tutor.

Subtraction

I can subtract using the number four.

READ

In Lesson XXVIII, we are learning to subtract, or remove, **four from** a first number. Some of the sentences are removing, or subtracting, some number so that the answer is **four**.

Practice saying each of these sentences at the top of page 31, using objects. You need to learn each of these sentences so go over them many times.

STUDY

4 from 9 leaves 5   5 from 9 leaves 4

RECITE

Write the numbers on the lines.

1. 4 from 4 leaves _____   **11.** 0 from 4 leaves _____

2. 4 from 5 leaves _____   **12.** 1 from 5 leaves _____

3. 4 from 6 leaves _____   **13.** 2 from 6 leaves _____

4. 4 from 7 leaves _____   **14.** 3 from 7 leaves _____

5. 4 from 8 leaves _____   **15.** 4 from 8 leaves _____

6. 4 from 9 leaves _____   **16.** 5 from 9 leaves _____

7. 4 from 10 leaves _____   **17.** 6 from 10 leaves _____

8. 4 from 11 leaves _____   **18.** 7 from 11 leaves _____

9. 4 from 12 leaves _____   **19.** 8 from 12 leaves _____

10. 4 from 13 leaves _____   **20.** 9 from 13 leaves _____

Answer the questions 1 through 9 in Lesson XXVIII, page 31 in your textbook. Ask your teacher/tutor to ask **you** the questions.

Subtraction

I can subtract using the number five.

READ

In Lesson XXIX, we are learning to subtract, or remove, **five from** a first number. Some of the sentences are removing, or subtracting, some number so that the answer is **five**.

Practice saying each of these sentences at the top of page 32, using objects. You need to learn each of these sentences so go over them many times.

STUDY

5 from 8 leaves 3

✳✳✳✳✳ * * *

3 from 8 leaves 5

✳✳✳ * * * *

RECITE

Write the numbers on the lines.

1. 5 from 5 leaves _____
2. 5 from 6 leaves _____
3. 5 from 7 leaves _____
4. 5 from 8 leaves _____
5. 5 from 9 leaves _____
6. 5 from 10 leaves _____
7. 5 from 11 leaves _____
8. 5 from 12 leaves _____
9. 5 from 13 leaves _____
10. 5 from 14 leaves _____

11. 0 from 5 leaves _____
12. 1 from 6 leaves _____
13. 2 from 7 leaves _____
14. 3 from 8 leaves _____
15. 4 from 9 leaves _____
16. 5 from 10 leaves _____
17. 6 from 11 leaves _____
18. 7 from 12 leaves _____
19. 8 from 13 leaves _____
20. 9 from 14 leaves _____

Answer the questions 1 through 9 in Lesson XXIX, page 32 in your textbook. Ask your teacher/tutor to ask **you** the questions.

Subtraction

I can subtract using the number six.

READ

In Lesson XXX, we are learning to subtract, or remove, **six from** a first number. Some of the sentences are removing, or subtracting, some number so that the answer is **six**.

Practice saying each of these sentences at the top of page 33, using objects. You need to learn each of these sentences so go over them many times.

STUDY

6 from 10 leaves 4   4 from 10 leaves 6

✳✳✳✳✳ * * * *   ✳✳✳✳ * * * * *

RECITE

Write the numbers on the lines.

1. 6 from 6 leaves _____	**11.** 0 from 6 leaves _____
2. 6 from 7 leaves _____	**12.** 1 from 7 leaves _____
3. 6 from 8 leaves _____	**13.** 2 from 8 leaves _____
4. 6 from 9 leaves _____	**14.** 3 from 9 leaves _____
5. 6 from 10 leaves _____	**15.** 4 from 10 leaves _____
6. 6 from 11 leaves _____	**16.** 5 from 11 leaves _____
7. 6 from 12 leaves _____	**17.** 6 from 12 leaves _____
8. 6 from 13 leaves _____	**18.** 7 from 13 leaves _____
9. 6 from 14 leaves _____	**19.** 8 from 14 leaves _____
10. 6 from 15 leaves _____	**20.** 9 from 15 leaves _____

Answer the questions 1 through 9 in Lesson XXX, page 33 in your textbook. Ask your teacher/tutor to ask **you** the questions.

Subtraction

I can subtract using the number seven.

READ

In Lesson XXXI, we are learning to subtract, or remove, **seven from** a first number. Some of the sentences are removing, or subtracting, some number so that the answer is **seven**.

Practice saying each of these sentences at the top of page 34, using objects. You need to learn each of these sentences so go over them many times.

STUDY

7 from 9 leaves 2

2 from 9 leaves 7

RECITE
Write the numbers on the lines.

1. 7 from 7 leaves _____

2. 7 from 8 leaves _____

3. 7 from 9 leaves _____

4. 7 from 10 leaves _____

5. 7 from 11 leaves _____

6. 7 from 12 leaves _____

7. 7 from 13 leaves _____

8. 7 from 14 leaves _____

9. 7 from 15 leaves _____

10. 7 from 16 leaves _____

11. 0 from 7 leaves _____

12. 1 from 8 leaves _____

13. 2 from 9 leaves _____

14. 3 from 10 leaves _____

15. 4 from 11 leaves _____

16. 5 from 12 leaves _____

17. 6 from 13 leaves _____

18. 7 from 14 leaves _____

19. 8 from 15 leaves _____

20. 9 from 16 leaves _____

Answer the questions 1 through 9 in Lesson XXXI, page 34 in your textbook. Ask your teacher/tutor to ask **you** the questions.

Review II

Remember that when we **subtract**, we **remove** some objects **from** the first number, the big group. We then ask how many we have **left**. The words **from** and **leaves** are used in subtraction sentences.

When we subtract **four from** a number, we remove **four from** the big number. This **leaves four less** in the group, the answer.

When we subtract **five from** a number, we remove **five from** the big number. This **leaves five less** in the group, the answer.

When we subtract **six from** a number, we remove **six from** the big number. This **leaves six less** in the group, the answer.

When we subtract **seven from** a number, we remove **seven from** the big number. This **leaves seven less** in the group, the answer.

Write the numbers on the lines.

1. 5 from 5 leaves _____
2. 5 from 6 leaves _____
3. 5 from 7 leaves _____
4. 5 from 8 leaves _____
5. 5 from 9 leaves _____
6. 5 from 10 leaves _____
7. 5 from 11 leaves _____
8. 5 from 12 leaves _____
9. 5 from 13 leaves _____
10. 5 from 14 leaves _____
11. 6 from 6 leaves _____
12. 6 from 7 leaves _____
13. 6 from 8 leaves _____
14. 6 from 9 leaves _____
15. 6 from 10 leaves _____

16. 6 from 11 leaves _____
17. 6 from 12 leaves _____
18. 6 from 13 leaves _____
19. 6 from 14 leaves _____
20. 6 from 15 leaves _____
21. 7 from 7 leaves _____
22. 7 from 8 leaves _____
23. 7 from 9 leaves _____
24. 7 from 10 leaves _____
25. 7 from 11 leaves _____
26. 7 from 12 leaves _____
27. 7 from 13 leaves _____
28. 7 from 14 leaves _____
29. 7 from 15 leaves _____
30. 7 from 16 leaves _____

Quiz II

Write the numbers on the lines.

1. 4 from 11 leaves _____   **11.** 5 from 9 leaves _____

2. 6 from 9 leaves _____   **12.** 6 from 15 leaves _____

3. 4 from 6 leaves _____   **13.** 4 from 7 leaves _____

4. 5 from 10 leaves _____   **14.** 7 from 12 leaves _____

5. 6 from 13 leaves _____   **15.** 6 from 8 leaves _____

6. 4 from 9 leaves _____   **16.** 5 from 12 leaves _____

7. 7 from 13 leaves _____   **17.** 4 from 8 leaves _____

8. 5 from 14 leaves _____   **18.** 7 from 14 leaves _____

9. 4 from 5 leaves _____   **19.** 6 from 10 leaves _____

10. 6 from 12 leaves _____   **20.** 7 from 16 leaves _____

21. Four of seven ducks flew away. _____ ducks were left.

22. Eight oranges taken from twelve oranges leaves _____ oranges.

23. Mary spent 6¢ of 8¢. _____¢ left.

24. Samuel lost seven of twelve marbles. He had _____ marbles left.

25. Seven lemons from thirteen lemons leaves _____ lemons.

Answer five oral questions from Lessons XXVIII, XXIX, XXX, or XXXI. These questions should be asked by a teacher/tutor.

Subtraction

I can subtract using the number eight.

READ

In Lesson XXXII, we are learning to subtract, or remove, **eight from** a first number. Some of the sentences are removing, or subtracting, some number so that the answer is **eight**.

Practice saying each of these sentences at the top of page 35, using objects. You need to learn each of these sentences so go over them many times.

STUDY

8 from 12 leaves 4

✻✻✻✻ ✱ ✱
✻✻✻✻ ✱ ✱

4 from 12 leaves 8

✻✻ ✱ ✱ ✱ ✱
✻✻ ✱ ✱ ✱ ✱

RECITE

Write the numbers on the lines.

1. 8 from 8 leaves _____
2. 8 from 9 leaves _____
3. 8 from 10 leaves _____
4. 8 from 11 leaves _____
5. 8 from 12 leaves _____
6. 8 from 13 leaves _____
7. 8 from 14 leaves _____
8. 8 from 15 leaves _____
9. 8 from 16 leaves _____
10. 8 from 17 leaves _____

11. 0 from 8 leaves _____
12. 1 from 9 leaves _____
13. 2 from 10 leaves _____
14. 3 from 11 leaves _____
15. 4 from 12 leaves _____
16. 5 from 13 leaves _____
17. 6 from 14 leaves _____
18. 7 from 15 leaves _____
19. 8 from 16 leaves _____
20. 9 from 17 leaves _____

Answer the questions 1 through 9 in Lesson XXXII, page 35 in your textbook. Ask your teacher/tutor to ask **you** the questions.

Subtraction

I can subtract using the number nine.

READ

In Lesson XXXIII, we are learning to subtract, or remove, **nine from** a first number. Some of the sentences are removing, or subtracting, some number so that the answer is **nine**.

Practice saying each of these sentences at the top of page 36, using objects. You need to learn each of these sentences so go over them many times.

STUDY

9 from 11 leaves 2

✳ ✳ ✳ ✳ ✳ ✳
✳ ✳ ✳ ✳

2 from 11 leaves 9

✳ ✳ ✳ ✳ ✳ ✳
✳ ✳ ✳ ✳

RECITE

Write the numbers on the lines.

1. 9 from 9 leaves _____

2. 9 from 10 leaves _____

3. 9 from 11 leaves _____

4. 9 from 12 leaves _____

5. 9 from 13 leaves _____

6. 9 from 14 leaves _____

7. 9 from 15 leaves _____

8. 9 from 16 leaves _____

9. 9 from 17 leaves _____

10. 9 from 18 leaves _____

11. 0 from 9 leaves _____

12. 1 from 10 leaves _____

13. 2 from 11 leaves _____

14. 3 from 12 leaves _____

15. 4 from 13 leaves _____

16. 5 from 14 leaves _____

17. 6 from 15 leaves _____

18. 7 from 16 leaves _____

19. 8 from 17 leaves _____

20. 9 from 18 leaves _____

Answer the questions 1 through 9 in Lesson XXXIII, page 36 in your textbook. Ask your teacher/tutor to ask **you** the questions.

Subtraction

I can subtract using the number ten.

READ

In Lesson XXXIV, we are learning to subtract, or remove, **ten from** a first number. Some of the sentences are removing, or subtracting, some number so that the answer is **ten**.

Practice saying each of these sentences at the top of page 37, using objects. You need to learn each of these sentences so go over them many times.

STUDY

10 from 14 leaves 4

✳ ✳ ✳ ✳ * *
✳ ✳ ✳ ✳ * *

4 from 14 leaves 10

✳ ✳ * * * *
✳ ✳ * * * *

RECITE

Write the numbers on the lines.

1. 10 from 10 leaves _____

2. 10 from 11 leaves _____

3. 10 from 12 leaves _____

4. 10 from 13 leaves _____

5. 10 from 14 leaves _____

6. 10 from 15 leaves _____

7. 10 from 16 leaves _____

8. 10 from 17 leaves _____

9. 10 from 18 leaves _____

10. 10 from 19 leaves _____

11. 0 from 10 leaves _____

12. 1 from 11 leaves _____

13. 2 from 12 leaves _____

14. 3 from 13 leaves _____

15. 4 from 14 leaves _____

16. 5 from 15 leaves _____

17. 6 from 16 leaves _____

18. 7 from 17 leaves _____

19. 8 from 18 leaves _____

20. 9 from 19 leaves _____

Answer the questions 1 through 9 in Lesson XXXIV, page 37 in your textbook. Ask your teacher/tutor to ask **you** the questions.

Story Problems

I can work subtraction story problems.

READ

All of the subtraction lessons have had story problems. We must read the English words and then set up a subtraction problem. After we state the problem, then we find the answer.

STUDY

Margaret had 10 cookies.
She gave 8 of them away.
Now Margaret has __2__ cookies.

8 cookies from 10 cookies leaves 2 cookies

2 cookies

Oliver is 4 years old.
Jane is 11 years old.
Jane is __7__ years older than Oliver.

4 years from 11 years leaves _____ years.

7 years

RECITE
Work these story problems.

1. Eliza had 6 birds in a cage.
   She let 2 of them out.
   Now Eliza has ____ birds.

2. Mary had 8 cents.
   She spent 6 cents.
   Now Mary has ____ cents.

3. Henry bought 15 pens.
   He lost 7 pens.
   Henry now has ____ pens.

4. Jane has 7 pencils.
   How many more pencils
   does she need to have 15? ____

5. George had 16 marbles.
   He lost 9 of them.
   George now has ____ marbles.

6. I owe 17 dollars.
   If I pay all but 7,
   how many will I pay? ____

17 (seventeen)

Review III

When we subtract, we remove objects from the first number, the big group. We then ask ourselves how many we have left. The words **from** and **leaves** are used in subtraction sentences.

When we subtract **eight from** a number, we remove **eight from** the big number. This **leaves eight less** in the group, the answer.

When we subtract **nine from** a number, we remove **nine from** the big number. This **leaves nine less** in the group, the answer.

When we subtract **ten from** a number, we remove **ten from** the big number. This **leaves ten less** in the group, the answer.

Story problems are problems written in English words. We must first read the problem, set it up, and then find the answer.

Write the numbers on the lines.

1. 8 from 8 leaves _____
2. 8 from 9 leaves _____
3. 8 from 10 leaves _____
4. 8 from 11 leaves _____
5. 8 from 12 leaves _____
6. 8 from 13 leaves _____
7. 8 from 14 leaves _____
8. 8 from 15 leaves _____
9. 8 from 16 leaves _____
10. 8 from 17 leaves _____
11. 9 from 9 leaves _____
12. 9 from 10 leaves _____
13. 9 from 11 leaves _____
14. 9 from 12 leaves _____
15. 9 from 13 leaves _____
16. 9 from 14 leaves _____
17. 9 from 15 leaves _____
18. 9 from 16 leaves _____
19. 9 from 17 leaves _____
20. 9 from 18 leaves _____
21. 10 from 10 leaves _____
22. 10 from 11 leaves _____
23. 10 from 12 leaves _____
24. 10 from 13 leaves _____
25. 10 from 14 leaves _____
26. 10 from 15 leaves _____
27. 10 from 16 leaves _____
28. 10 from 17 leaves _____
29. 10 from 18 leaves _____
30. 10 from 19 leaves _____

Quiz III

Write the numbers on the lines.

1. 8 from 13 leaves _____   11. 9 from 16 leaves _____

2. 10 from 16 leaves _____   12. 8 from 12 leaves _____

3. 9 from 12 leaves _____   13. 10 from 19 leaves _____

4. 9 from 17 leaves _____   14. 6 from 15 leaves _____

5. 4 from 14 leaves _____   15. 9 from 18 leaves _____

6. 10 from 15 leaves _____   16. 10 from 18 leaves _____

7. 8 from 17 leaves _____   17. 8 from 18 leaves _____

8. 9 from 10 leaves _____   18. 5 from 15 leaves _____

9. 8 from 14 leaves _____   19. 3 from 12 leaves _____

10. 3 from 13 leaves _____   20. 5 from 13 leaves _____

21. Ten dollars from twelve dollars leaves _____ dollars.

22. Ten trees of fourteen trees cut down leaving _____ trees.

23. Four birds out of thirteen birds died leaving _____ birds.

24. I bought a kite for nine cents and sold it for eighteen cents making

_____ cents.

25. Eight people out of sixteen people leave the room leaving _____ people.

Answer five oral questions from Lessons XXXII through XXXIV. These questions should be asked by a teacher/tutor.

How Many Less?

I can subtract to find how many less.

READ

The word **from** tells us to subtract. We subract the smaller number **from** the larger number.

Another word that tells us to subtract is **less**. When we subtract, we are removing objects. The number of objects is **less** the number being removed. The group is getting smaller, or **less**.

When we subtract, we are thinking in reverse of addition. The numbers are getting smaller in value, or **less**. In addition, numbers are getting larger in value, or **more**.

STUDY

How many are 4 less 2.   Remove 2

The answer is 4 less 2 are 2.

RECITE

Write the numbers on the lines.

1. 10 less 6 are _____   **11.** 16 less 7 are _____

2. 6 less 3 are _____   **12.** 15 less 10 are _____

3. 9 less 3 are _____   **13.** 17 less 10 are _____

4. 11 less 4 are _____   **14.** 8 less 3 are _____

5. 12 less 8 are _____   **15.** 11 less 9 are _____

6. 19 less 9 are _____   **16.** 7 less 2 are _____

7. 13 less 6 are _____   **17.** 9 less 5 are _____

8. 13 less 3 are _____   **18.** 12 less 3 are _____

9. 14 less 7 are _____   **19.** 17 less 9 are _____

10. 7 less 4 are _____   **20.** 7 less 5 are _____

Practice working problems 1 through 10 in Lesson XXXV, page 38.
Ask your teacher/tutor to ask **you** the questions.

How Many Less?

| I can subtract to find how many less. |

READ

Remember that when we add, the numbers are getting larger in value, or more. When we subtract, the numbers are getting smaller in value, or less.

To find how many less, we remove the number less from the first number. The answer is the number for the first number less the second number.

STUDY

How many are 8 less 6.

The answer is 8 less 6 are 2.

RECITE
Write the numbers on the lines.

1. 10 less 7 are _____
2. 13 less 10 are _____
3. 13 less 7 are _____
4. 18 less 9 are _____
5. 14 less 8 are _____
6. 12 less 5 are _____
7. 13 less 5 are _____
8. 15 less 5 are _____
9. 19 less 10 are _____
10. 9 less 2 are _____

11. 13 less 4 are _____
12. 11 less 5 are _____
13. 16 less 8 are _____
14. 12 less 10 are _____
15. 15 less 8 are _____
16. 12 less 6 are _____
17. 9 less 4 are _____
18. 15 less 7 are _____
19. 12 less 7 are _____
20. 17 less 7 are _____

Practice working problems 11 through 20 in Lesson XXXV, page 38.
Ask your teacher/tutor to ask **you** the questions.

How Many Less?

I can subtract to find how many less?

READ

Remember that when we subtract, we are counting in **reverse**. When we add, we count **forward**. When we subtract, we count **backward**.

To find how many less, we remove the number less from the first number. We can find the answer by counting in reverse the number less from the first number.

STUDY

How many are 5 less 2?

Count backward 2 less from 5.

⑤ ④ 3 2 1

The answer is 3.

RECITE
Write the numbers on the lines.

1. 10 less 3 are _____   **11.** 14 less 5 are _____

2. 14 less 6 are _____   **12.** 11 less 2 are _____

3. 5 less 3 are _____   **13.** 8 less 4 are _____

4. 10 less 8 are _____   **14.** 6 less 2 are _____

5. 14 less 4 are _____   **15.** 13 less 9 are _____

6. 9 less 6 are _____   **16.** 18 less 8 are _____

7. 12 less 2 are _____   **17.** 10 less 2 are _____

8. 13 less 8 are _____   **18.** 16 less 6 are _____

9. 10 less 5 are _____   **19.** 6 less 4 are _____

10. 9 less 7 are _____   **20.** 8 less 2 are _____

Practice working problems 21 through 27 in Lesson XXXV, page 38.
Ask your teacher/tutor to ask **you** the questions.

Addition and Subtraction

I can do addition and subtraction.

READ

Some problems combine addition and subtraction. The word **and** tells us to add the numbers. The word **less** tells us to subtract the numbers.

We are working two problems. First we are adding. Second we are subtracting.

STUDY

How many are 3 and 3 less 4?

3 and 3 are 6

6 less 4 are 2

Count forward for adding.
three, <u>four</u>, <u>five</u>, <u>six</u>

Count backward for subtracting.
six, <u>five</u>, <u>four</u>, <u>three</u>, <u>two</u>

RECITE
Write the numbers on the lines.

1. 2 and 5 less 3 are _____

2. 8 and 9 less 7 are _____

3. 7 and 10 less 8 are _____

4. 3 and 4 less 5 are _____

5. 2 and 6 less 3 are _____

6. 7 and 9 less 6 are _____

7. 3 and 5 less 6 are _____

8. 6 and 10 less 9 are _____

9. 8 and 10 less 9 are _____

10. 9 and 8 less 10 are _____

11. 10 and 7 less 9 are _____

12. 5 and 2 less 4 are _____

13. 5 and 3 less 4 are _____

14. 10 and 6 less 8 are _____

15. 4 and 4 less 6 are _____

16. 10 and 6 less 7 are _____

17. 9 and 9 less 10 are _____

18. 3 and 5 less 2 are _____

19. 6 and 7 less 4 are _____

20. 5 and 8 less 6 are _____

Practice working problems 1 through 10 in Lesson XXXVI, page 39.
Ask your teacher/tutor to ask **you** the questions.

Review IV

The word **from** tells us to subtract. We subtract the smaller number **from** the larger number.

Another word that tells us to subtract is **less**. When we subtract, we are removing objects The number of objects is **less** the number being removed. The group of objects is getting smaller, or **less**.

When we subtract, we are counting in reverse. Addition is counting forward, or more. Subtraction is counting backward, or less.

The answer for addition is a larger number. The answer for subtraction is a smaller number.

In some problems, we can do both addition and subtraction. The word **and** tells us to add the numbers. The word **less** tells us to subtract the numbers.

Write the numbers on the lines.

1. 12 less 5 are _____   **6.** 18 less 10 are _____

2. 14 less 6 are _____   **7.** 17 less 8 are _____

3. 10 less 2 are _____   **8.** 16 less 7 are _____

4. 15 less 8 are _____   **9.** 14 less 8 are _____

5. 12 less 3 are _____   **10.** 15 less 7 are _____

11. 5 and 3 less 4 are _____

12. 2 and 6 less 3 are _____

13. 6 and 10 less 9 are _____

14. 9 and 8 less 10 are _____

15. 6 and 7 less 4 are _____

Quiz IV

Write the numbers on the lines.

1. 11 less 9 are _____

2. 19 less 9 are _____

3. 14 less 7 are _____

4. 12 less 3 are _____

5. 13 less 7 are _____

6. 15 less 8 are _____

7. 9 less 4 are _____

8. 17 less 7 are _____

9. 9 less 2 are _____

10. 11 less 5 are _____

11. 10 less 3 are _____

12. 6 less 2 are _____

13. 9 less 3 are _____

14. 12 less 2 are _____

15. 10 less 5 are _____

16. 18 less 8 are _____

17. 14 less 6 are _____

18. 13 less 9 are _____

19. 15 less 6 are _____

20. 17 less 8 are _____

21. 5 and 3 less 4 are _____

22. 6 and 7 less 5 are _____

23. 5 and 8 less 6 are _____

24. 10 and 7 less 9 are _____

25. 6 and 10 less 8 are _____

Answer five oral questions from Lessons XXV through XXXIV. These questions should be asked by a teacher/tutor.

Addition and Subtraction

I can do addition and subtraction.

READ

Remember that we count forward when we add. We count backward in reverse when we subtract.

We can work two problems in one.

First we add the numbers. Second we subtract the numbers. Remember the word **and** tells us to add. The word **less** tells us to subtract.

STUDY

How many are 3 and 9 less 10?

3 and 9 are 12.

12 less 10 are 2.

Count forward for adding.
three, <u>four</u>, <u>five</u>, <u>six</u>, <u>seven</u>,
<u>eight</u>, <u>nine</u>, <u>ten</u>, <u>eleven</u>, <u>twelve</u>

Count backward for subtracting.
twelve, <u>eleven</u>, <u>ten</u>, <u>nine</u>, <u>eight</u>,
<u>seven</u>, <u>six</u>, <u>five</u>, <u>four</u>, <u>three</u>, <u>two</u>

RECITE

Write the numbers on the lines.

1. 7 and 7 less 5 are _____

2. 9 and 3 less 7 are _____

3. 8 and 5 less 6 are _____

4. 4 and 8 less 9 are _____

5. 6 and 7 less 3 are _____

6. 5 and 9 less 7 are _____

7. 7 and 8 less 5 are _____

8. 5 and 6 less 9 are _____

9. 2 and 7 less 4 are _____

10. 6 and 6 less 4 are _____

11. 10 and 5 less 7 are _____

12. 3 and 8 less 5 are _____

13. 8 and 7 less 6 are _____

14. 5 and 5 less 2 are _____

15. 8 and 6 less 4 are _____

16. 8 and 6 less 9 are _____

17. 7 and 5 less 8 are _____

18. 3 and 7 less 4 are _____

19. 6 and 8 less 10 are _____

20. 6 and 5 less 4 are _____

Practice working problems 11 through 20 in Lesson XXXVI, page 39.
Ask your teacher/tutor to ask **you** the questions.

Addition and Subtraction

I can do addition and subtraction.

READ

Look for the words **and** and **less**. The word **and** tells us to add. The word **less** tells us to subtract. We do addition first. After we add the numbers, then we subtract.

When we add, the numbers get larger in value, or **more**. When we subtract, the numbers get smaller in value or **less**.

STUDY

How many are 4 and 9 less 5?

4 and 9 are 13.

13 less 5 are 8.

RECITE

Write the numbers on the lines.

1. 6 and 3 less 7 are _____

2. 5 and 7 less 3 are _____

3. 10 and 2 less 5 are _____

4. 7 and 4 less 8 are _____

5. 7 and 6 less 10 are _____

6. 9 and 6 less 7 are _____

7. 2 and 8 less 5 are _____

8. 5 and 4 less 2 are _____

9. 8 and 6 less 7 are _____

10. 8 and 8 less 10 are _____

11. 5 and 9 less 7 are _____

12. 3 and 7 less 4 are _____

13. 8 and 4 less 7 are _____

14. 2 and 9 less 5 are _____

15. 9 and 5 less 8 are _____

16. 3 and 8 less 2 are _____

17. 1 and 9 less 6 are _____

18. 5 and 8 less 6 are _____

19. 7 and 7 less 5 are _____

20. 5 and 7 less 8 are _____

Practice working problems 21 through 27 in Lesson XXXVI, page 39. Ask your teacher/tutor to ask **you** the questions.

Story Problems

I can work problems stated in words.

READ

We call problems written in English words story problems. Some problems, such as problems one through five in Lesson XXXVII are not stories.

When problems are written in words, we must carefully read the words. After we read the words, we write the problem in numbers. Then we find the answer.

STUDY

Begin with 20 and subtract by 2's to 0.

20, 18, 16, 14, 12, 10, 8, 6, 4, 2, 0

RECITE

Write the answers for problems 1 through 5 in Lesson XXXVII, page 40 in your textbook.

1. 20, ____ , ____ , ____ , ____ , ____ , ____ , ____ , ____ , ____ , ____

2. 18, ____ , ____ , ____ , ____ , ____ , ____

3. 20, ____ , ____ , ____ , ____ , ____

4. 20, ____ , ____ , ____ , ____

5. 18, ____ , ____ , ____

Story Problems

I can work story problems.

READ

Story problems are problems written in English words. The first thing we must do is carefully read the problem. Next we write the problem in numbers and then we find the answer.

STUDY

Francis has 10 cents in two pockets.
There are 4 cents in one.
How many in the other?

Problem: 10 cents less 4 cents are _____ cents.
Answer: 6 cents

RECITE

Work these story problems.

1. I think of two numbers that
together make 8.
One number is 5.

   What is the other? _____

2. Mary had 11 apples.
She gave 4 to Lucy and
5 to Nancy.

   How many had she left? _____

3. I have 10 cents in one hand
and 5 cents in the other.
If I take 3 cents from
each hand, how many
cents will I have in both

   hands? _____

4. Albert bought 10 apples.
He sold 3 and ate 2.
How many apples had

   he left? _____

Work all problems 6 through 14 in Lesson XXXVII, page 40.

Review V

Some problems combine both addition and subtraction. The word **and** tells us to add. The word **less** tells us to subtract. In the problems on page 39, we add first and then subtract second.

When we add, we count forward. The number of the answer is a larger number, or more. When we subtract, we count backward. The number of the answer is a smaller number, or less. The counting for subtraction is reverse to the counting for addition.

Many times problems are written in English words. Some problems are story problems, or the problem is based on a story.

When the problems are written in words, we must read carefully. Then we write the problem in numbers. Finally we find the answer to the problem.

Write the numbers on the lines.

1. 6 and 3 less 7 are _____

2. 5 and 4 less 2 are _____

3. 5 and 7 less 3 are _____

4. 3 and 8 less 2 are _____

5. 7 and 8 less 5 are _____

6. 9 and 5 less 8 are _____

7. 8 and 8 less 10 are _____

8. 2 and 8 less 5 are _____

9. 5 and 7 less 8 are _____

10. 8 and 6 less 9 are _____

Work these story problems.

11. I have 11 cents. In one pocket, I have 5 cents. How much in the other pocket? ____

12. James had 8 marbles and John had 7 marbles. Henry had 10 less marbles than James and John together. How many marbles had Henry? ____

Quiz V

Write the numbers on the lines.

1. 8 and 7 less 6 are _____

2. 7 and 5 less 8 are _____

3. 5 and 6 less 9 are _____

4. 10 and 5 less 7 are _____

5. 4 and 8 less 9 are _____

6. 5 and 5 less 2 are _____

7. 9 and 6 less 7 are _____

8. 5 and 4 less 2 are _____

9. 3 and 7 less 4 are _____

10. 1 and 9 less 6 are _____

Write the answers for problems 1 through 5 in Lesson XXXVII, page 40 in your textbook.

11. 20, ____ , ____ , ____ , ____ , ____ , ____ , ____ , ____ , ____ , ____

12. 18, ____ , ____ , ____ , ____ , ____ , ____

13. 20, ____ , ____ , ____ , ____ , ____

14. 20, ____ , ____ , ____ , ____

15. 18, ____ , ____ , ____

Work these story problems.

16. Mary had 11 apples.
    She gave 4 to Lucy
    and 5 to Nancy.

    How many had she left? ____

17. I think of two numbers
    that together make 13.
    One number is 5.

    What is the other? ____

Working with Groups

I can think of numbers in groups.

READ

When we add or subtract, we think of single objects. We add objects by combining objects and we subtract objects by removing objects.

We can also add groups of objects.

Think about four bird nests. Each nest has three eggs. Each nest is a group. Each group has three objects in it. We find the number of eggs by saying "Four groups of three eggs are twelve eggs."

STUDY

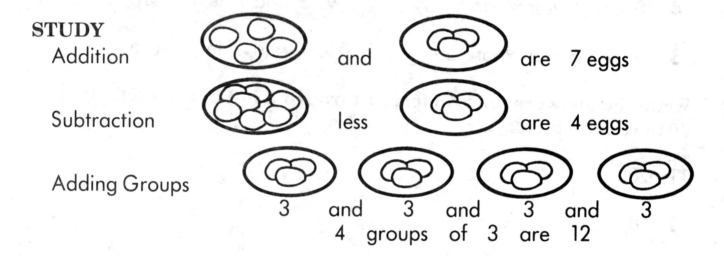

Addition   and   are   7 eggs

Subtraction   less   are   4 eggs

Adding Groups

3   and   3   and   3   and   3

4   groups   of   3   are   12

RECITE

Write the numbers on the lines.

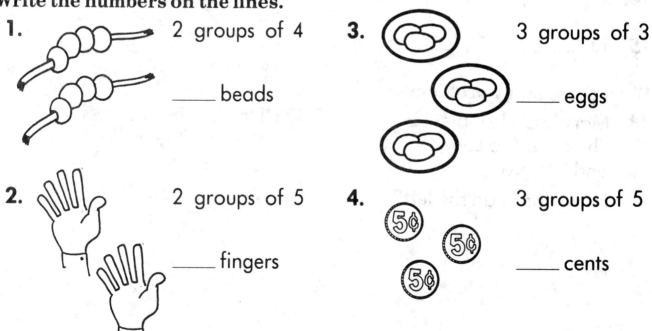

1.   2 groups of 4

_____ beads

2.   2 groups of 5

_____ fingers

3.   3 groups of 3

_____ eggs

4.   3 groups of 5

_____ cents

Ray's New Primary Arithmetic Workbook

Book Two Test

Name _____

Date _____

Possible score _____50_____

My score _____

_____ x 2 = _____%
score

Book Two Test

Write the numbers on the lines.

1. 3 from 9 leaves _____
2. 4 from 11 leaves _____
3. 6 from 12 leaves _____
4. 5 from 13 leaves _____
5. 6 from 14 leaves _____
6. 8 from 15 leaves _____
7. 7 from 16 leaves _____

8. 10 less 6 are _____
9. 15 less 9 are _____
10. 13 less 7 are _____
11. 11 less 2 are _____
12. 9 less 5 are _____
13. 14 less 9 are _____
14. 18 less 10 are _____

15. Three of seven ducks flew away. _____ ducks left.

16. Jon spent 10¢ of 13¢. _____¢ left.

17. Eight apples from twelve apples leaves _____ apples.

18. Jack lost nine of sixteen marbles. He has _____ marbles left.

19. 6 and 4 less 7 are _____
20. 9 and 5 less 6 are _____
21. 7 and 8 less 9 are _____
22. 8 and 4 less 5 are _____
23. 9 and 7 less 8 are _____

24. 4 times 3 are _____
25. 5 times 2 are _____
26. 2 times 3 are _____
27. 1 time 6 is _____
28. 3 times 6 are _____
29. 8 times 4 are _____
30. 6 times 2 are _____

31. 5 times 5 are _____
32. 8 times 3 are _____
33. 9 times 4 are _____
34. 2 times 9 are _____
35. 3 times 5 are _____
36. 3 times 9 are _____
37. 7 times 3 are _____

Primary Arithmetic
Book 2
Answer Key

Page 2
1. 3
2. 3
3. 6

Page 3
1. 0
2. 1
3. 2
4. 3
5. 4
6. 5
7. 6
8. 7
9. 8
10. 9
11. 1
12. 1
13. 1
14. 1
15. 1
16. 1
17. 1
18. 1
19. 1
20. 1

Page 4
1. 0
2. 1
3. 2
4. 3
5. 4
6. 5
7. 6
8. 7
9. 8
10. 9
11. 2
12. 2
13. 2
14. 2
15. 2
16. 2
17. 2
18. 2
19. 2
20. 2

Page 5
1. 0
2. 1
3. 2
4. 3
5. 4
6. 5
7. 6
8. 7
9. 8
10. 9
11. 3
12. 3
13. 3
14. 3
15. 3
16. 3
17. 3
18. 3
19. 3
20. 3

Page 6
1. 0
2. 1
3. 2
4. 3
5. 4
6. 5
7. 6
8. 7
9. 8
10. 9
11. 0
12. 1
13. 2
14. 3
15. 4
16. 5
17. 6
18. 7
19. 8
20. 9
21. 0
22. 1
23. 2
24. 3
25. 4
26. 5
27. 6
28. 7
29. 8
30. 9

Page 8
1. 0
2. 1
3. 2
4. 3
5. 4
6. 5
7. 6
8. 7
9. 8
10. 9
11. 4
12. 4
13. 4
14. 4
15. 4
16. 4
17. 4
18. 4
19. 4
20. 4

Page 9
1. 0
2. 1
3. 2
4. 3
5. 4
6. 5
7. 6
8. 7
9. 8
10. 9
11. 5
12. 5
13. 5
14. 5
15. 5
16. 5
17. 5
18. 5
19. 5
20. 5

Page 10
1. 0
2. 1
3. 2
4. 3
5. 4
6. 5
7. 6
8. 7
9. 8
10. 9
11. 6
12. 6
13. 6
14. 6
15. 6
16. 6
17. 6
18. 6
19. 6
20. 6

Page 11
1. 0
2. 1
3. 2
4. 3
5. 4
6. 5
7. 6
8. 7
9. 8
10. 9
11. 7
12. 7
13. 7
14. 7
15. 7
16. 7
17. 7
18. 7
19. 7
20. 7

Page 12
1. 0
2. 1
3. 2
4. 3
5. 4
6. 5
7. 6
8. 7
9. 8
10. 9
11. 0
12. 1
13. 2
14. 3
15. 4
16. 5
17. 6
18. 7
19. 8
20. 9
21. 0
22. 1
23. 2
24. 3
25. 4
26. 5
27. 6
28. 7
29. 8
30. 9

Page 14
1. 0
2. 1
3. 2
4. 3
5. 4
6. 5
7. 6
8. 7
9. 8
10. 9
11. 8
12. 8
13. 8
14. 8
15. 8
16. 8
17. 8
18. 8
19. 8
20. 8

Page 15
1. 0
2. 1
3. 2
4. 3
5. 4
6. 5
7. 6
8. 7
9. 8
10. 9
11. 9
12. 9
13. 9
14. 9
15. 9
16. 9
17. 9
18. 9
19. 9
20. 9

Page 16
1. 0
2. 1
3. 2
4. 3
5. 4
6. 5
7. 6
8. 7
9. 8
10. 9
11. 10
12. 10
13. 10
14. 10
15. 10
16. 10
17. 10
18. 10
19. 10
20. 10

Page 17
1. 4
2. 2
3. 8
4. 8
5. 7
6. 10

Page 18
1. 0
2. 1
3. 2
4. 3
5. 4
6. 5
7. 6
8. 7
9. 8
10. 9
11. 0
12. 1
13. 2
14. 3
15. 4
16. 5

Answer Key (continued)

17. 6
18. 7
19. 8
20. 9
21. 0
22. 1
23. 2
24. 3
25. 4
26. 5
27. 6
28. 7
29. 8
30. 9

Page 20
1. 4
2. 3
3. 6
4. 7
5. 4
6. 10
7. 7
8. 10
9. 7
10. 3
11. 9
12. 5
13. 7
14. 5
15. 2
16. 5
17. 4
18. 9
19. 8
20. 2

Page 21
1. 3
2. 3
3. 6
4. 9
5. 6
6. 7
7. 8
8. 10
9. 9
10. 7
11. 9
12. 6
13. 8
14. 2
15. 7
16. 6
17. 5
18. 8
19. 5
20. 10

Page 22
1. 7
2. 8
3. 2
4. 2

5. 10
6. 3
7. 10
8. 5
9. 5
10. 2
11. 9
12. 9
13. 4
14. 4
15. 4
16. 10
17. 10
18. 10
19. 2
20. 6

Page 23
1. 4
2. 10
3. 9
4. 2
5. 5
6. 10
7. 2
8. 7
9. 9
10. 7
11. 8
12. 3
13. 4
14. 8
15. 2
16. 9
17. 8
18. 6
19. 9
20. 7

Page 24
1. 7
2. 8
3. 8
4. 7
5. 9
6. 8
7. 9
8. 9
9. 6
10. 8
11. 4
12. 5
13. 7
14. 7
15. 9

Page 26
1. 9
2. 5
3. 7
4. 3
5. 10
6. 7
7. 10

8. 2
9. 5
10. 8
11. 8
12. 6
13. 9
14. 8
15. 10
16. 5
17. 4
18. 6
19. 4
20. 7

Page 27
1. 2
2. 9
3. 7
4. 3
5. 3
6. 8
7. 5
8. 7
9. 7
10. 6
11. 7
12. 6
13. 5
14. 6
15. 6
16. 9
17. 7
18. 7
19. 9
20. 4

Page 28
1. 18, 16, 14, 12, 10, 8, 6, 4, 2, 0
2. 15, 12, 9, 6, 3, 0
3. 16, 12, 8, 4, 0
4. 15, 10, 5, 0
5. 12, 6, 0

Page 29
1. 3
2. 2 apples
3. 9 cents
4. 5 apples

Page 30
1. 2
2. 7
3. 9
4. 9
5. 10
6. 6
7. 6
8. 5
9. 4
10. 5
11. 6 cents
12. 5 marbles

Page 32
1. 8
2. 10
3. 9
4. 15

Page 33
1. 12
2. 16
3. 16
4. 6

Page 34
1. 1
2. 2
3. 3
4. 4
5. 5
6. 6
7. 7
8. 8
9. 9
10. 10
11. 1
12. 2
13. 3
14. 4
15. 5
16. 6
17. 7
18. 8
19. 9
20. 10

Page 35
1. 2 cents
2. 3 cents
3. 4 cents
4. 5 cents
5. $6
6. 7 miles
7. 8 cents
8. 9 cents
9. 10 cents

Page 36
1. 1
2. 2
3. 3
4. 4
5. 5
6. 6
7. 7
8. 8
9. 9
10. 10
11. 1
12. 2
13. 3
14. 4
15. 5
16. 6
17. 7
18. 8
19. 9
20. 10

Page 38
1. 2
2. 4
3. 6
4. 8
5. 10
6. 12
7. 14
8. 16
9. 18
10. 20
11. 2
12. 4
13. 6
14. 8
15. 10
16. 12
17. 14
18. 16
19. 18
20. 20

Page 39
1. 4 cents
2. 6 cents
3. 8 cents
4. 10 cents
5. 12 cents
6. 14 cents
7. 16 cents
8. 18 cents
9. 20 cents

Page 40
1. 3
2. 6
3. 9
4. 12
5. 15
6. 18
7. 21
8. 24
9. 27
10. 30
11. 3
12. 6
13. 9
14. 12
15. 15
16. 18
17. 21
18. 24
19. 27
20. 30

Page 41
1. 6 cents
2. 9 cents
3. 12 miles
4. 15 apples
5. 18 plums
6. 21 cents
7. 24 cents
8. 27 cents
9. 30 cents

Answer Key (continued)

Page 42
1. 4
2. 6
3. 8
4. 10
5. 12
6. 14
7. 16
8. 18
9. 20
10. 6
11. 9
12. 12
13. 15
14. 18
15. 21
16. 24
17. 27
18. 30
19. 8

20. 10
21. 12
22. 14
23. 16
24. 18
25. 12
26. 15
27. 18
28. 21
29. 24
30. 27

Page 44
1. 4
2. 8
3. 12
4. 16
5. 20
6. 24
7. 28

8. 32
9. 36
10. 40
11. 4
12. 8
13. 12
14. 16
15. 20
16. 24
17. 28
18. 32
19. 36
20. 40

Page 45
1. 8 feet
2. 12 pigeons
3. 16 cents
4. 20 cents
5. 24 quarters

6. 28 cents
7. 32 cents
8. 36 cents
9. 40 cents

Page 46
1. 5
2. 10
3. 15
4. 20
5. 25
6. 30
7. 35
8. 40
9. 45
10. 50
11. 5
12. 10
13. 15
14. 20

- -

Primary Arithmetic
Book 2
Quizzes and Tests

Quiz I
1. 7
2. 2
3. 7
4. 2
5. 3
6. 2
7. 7
8. 5
9. 1
10. 8
11. 4
12. 4
13. 1
14. 8
15. 6
16. 0
17. 9
18. 4
19. 6
20. 2
21. one
22. three
23. two
24. one
25. three

Quiz II
1. 7
2. 3
3. 2
4. 5
5. 7
6. 5
7. 6

8. 9
9. 1
10. 6
11. 4
12. 9
13. 3
14. 5
15. 2
16. 7
17. 4
18. 7
19. 4
20. 9
21. three
22. four
23. 2
24. five
25. six

Quiz III
1. 5
2. 6
3. 3
4. 8
5. 10
6. 5
7. 9
8. 1
9. 6
10. 10
11. 7
12. 4
13. 9
14. 9

15. 9
16. 8
17. 10
18. 10
19. 9
20. 8
21. two
22. four
23. nine
24. nine
25. eight

Quiz IV
1. 2
2. 10
3. 7
4. 9
5. 6
6. 7
7. 5
8. 10
9. 7
10. 6
11. 7
12. 4
13. 6
14. 10
15. 5
16. 10
17. 8
18. 4
19. 9
20. 9
21. 4

22. 8
23. 7
24. 8
25. 8

Quiz V
1. 9
2. 4
3. 2
4. 8
5. 3
6. 8
7. 8
8. 7
9. 6
10. 4
11. 18, 16, 14, 12,
    10, 8, 6, 4, 2, 0
12. 15, 12, 9, 6, 3, 0
13. 16, 12, 8, 4, 0
14. 15, 10, 5, 0
15. 12, 6, 0
16. 2 apples
17. 8

Quiz VI
1. 5
2. 6
3. 10
4. 4
5. 8
6. 6
7. 9
8. 2

9. 10
10. 7
11. 6
12. 4
13. 15
14. 8
15. 10

Quiz VII
1. 4
2. 12
3. 8
4. 6
5. 12
6. 14
7. 15
8. 16
9. 10
10. 9
11. 12
12. 18
13. 21
14. 24
15. 20
16. 18
17. 21
18. 16
19. 27
20. 30
21. 15
22. 8
23. 20
24. 24
25. 10

3(three)

Answer Key (continued)

15. 25
16. 30
17. 35
18. 40
19. 45
20. 50

Page 47
1. 10 cents
2. 15 miles
3. 20 plums

4. 25 chickens
5. 30 eggs
6. 35 cents
7. 40 cents
8. 45 cents
9. 50 cents

Page 48
1. 8
2. 12
3. 16
4. 20

5. 24
6. 28
7. 32
8. 36
9. 40
10. 10
11. 15
12. 20
13. 25
14. 30
15. 35
16. 40
17. 45

18. 50
19. 16
20. 20
21. 24
22. 28
23. 32
24. 36
25. 20
26. 25
27. 30
28. 35
29. 40
30. 45

Quizzes and Tests (continued)

Quiz VIII
1. 8
2. 20
3. 16
4. 12
5. 24
6. 28
7. 25
8. 32
9. 20
10. 15
11. 15
12. 36
13. 35
14. 40
15. 40
16. 36
17. 35
18. 32
19. 45
20. 50
21. 20
22. 30
23. 20
24. 28
25. 25

Quiz I-II
1. 7
2. 2
3. 8
4. 7
5. 1
6. 7
7. 2
8. 6
9. 8
10. 8

11. 3
12. 4
13. 8
14. 9
15. 8
16. 9
17. 6
18. 5
19. 7
20. 7
21. six
22. six
23. seven
24. nine
25. eight

Quiz III-IV
1. 5
2. 7
3. 10
4. 9
5. 9
6. 4
7. 9
8. 7
9. 9
10. 4
11. 10
12. 8
13. 6
14. 6
15. 6
16. 8
17. 4
18. 7
19. 6
20. 7

21. 6
22. 10
23. 5
24. 8
25. 8

Quiz V-VI
1. 7
2. 7
3. 3
4. 7
5. 8
6. 5
7. 7
8. 8
9. 7
10. 5
11. 4
12. 3
13. 2
14. 6
15. 10
16. 8
17. 9
18. 5
19. 10
20. 6
21. 12
22. 6
23. 10
24. 6
25. 9

Quiz VII-VIII
1. 8
2. 21
3. 10

4. 18
5. 32
6. 40
7. 27
8. 12
9. 35
10. 20
11. 36
12. 28
13. 25
14. 16
15. 24
16. 12
17. 9
18. 30
19. 4
20. 24
21. 15
22. 32
23. 30
24. 27
25. 45

Book Two Test
1. 6
2. 7
3. 6
4. 8
5. 8
6. 7
7. 9
8. 4
9. 6
10. 6
11. 9
12. 4
13. 5

14. 8
15. four
16. 3
17. four
18. seven
19. 3
20. 8
21. 6
22. 7
23. 8
24. 12
25. 10
26. 6
27. 6
28. 18
29. 32
30. 12
31. 25
32. 24
33. 36
34. 18
35. 15
36. 27
37. 21
38. 12
39. 28
40. 35
41. 12
42. 18, 16, 14, 12, 10, 8, 6, 4, 2
43. 16, 12, 8, 4
44. 15, 10, 5
45. 7
46. 6
47. 2
48. 18
49. 24 fish
50. 24 quarters

Write the numbers on the lines.

38. Six nests with two eggs in each nest are _____ eggs.

39. 7 boxes each having 4 pencils are _____ pencils.

40. Father has seven nickels in his pocket or _____ cents.

41. We have three strings of beads with four beads on each string or _____ beads.

42. Begin with 20 and subtract by 2's to 0.

20, ____ , ____ , ____ , ____ , ____ , ____ , ____ , ____ , ____ , 0

43. Begin with 20 and subtract by 4's to 0.

20, ____ , ____ , ____ , ____ , 0

44. Begin with 20 and subtract by 5's to 0.

20, ____ , ____ , ____ , 0

Work these word problems.

45. Oliver is 4 years old.
Jane is 11 years old.

Jane is ____ years
older than Oliver.

46. Samuel had 14 marbles.
He lost 8 of them.

Now Samuel has ____ marbles.

47. I had 12 dollars.
After spending part,
I had 10 dollars left.

I spent ____ dollars.

48. Mary had 3 hens.
Each hen had 6 chicks.

Mary had ____ chicks.

49. Each of 3 boys
caught 8 fish.
How many fish did
they all catch? ____

50. One apple has
4 quarters.
How many quarters
in 6 apples? ____

3 (three)

Working with Groups

I can think of numbers in groups.

READ

In Lesson XXXVIII, page 41, we are working with groups. We are asked about three groups of birds: the birds flying inside the shed, the birds at rest, and the birds flying outside the shed. Each group has the same number of birds. We can find how many birds by adding the three groups.

STUDY

How many birds are flying inside the shed?   3 birds
How many birds are at rest in the shed?   3 birds
How many birds are flying outside the shed?   3 birds

    3 groups of 3 birds are 9 birds.
    3 times 3 are 9.

RECITE

Write the numbers on the lines.

1. How many birds are
   three times four birds?
   How many are

   3 groups of 4? ____ birds

3. How many wings
   have eight swallows?
   How many are

   8 groups of 2? ____ wings

2. How many eggs are
   four times four eggs?
   How many are

   4 groups of 4? ____ eggs

4. How many are
   three times two?
   How many are

   three groups of 2? ____

Multiplication

I can learn multiplication by one.

READ

When we combine numbers in groups, we are doing multiplication. In Lesson XXXIX, page 42, we are working with groups of **one**, or **one** group of some number. Read the sentences at the top of page 42. Notice that there are two sentences with the same answer.

STUDY

1 time 4 is 4
1 group of 4 is 4

4 times 1 are 4
4 groups of 1 are 4

RECITE

Write the numbers on the lines.

1. 1 time 1 is _____

2. 1 time 2 is _____

3. 1 time 3 is _____

4. 1 time 4 is _____

5. 1 time 5 is _____

6. 1 time 6 is _____

7. 1 time 7 is _____

8. 1 time 8 is _____

9. 1 time 9 is _____

10. 1 time 10 is _____

11. 1 time 1 is _____

12. 2 times 1 are _____

13. 3 times 1 are _____

14. 4 times 1 are _____

15. 5 times 1 are _____

16. 6 times 1 are _____

17. 7 times 1 are _____

18. 8 times 1 are _____

19. 9 times 1 are _____

20. 10 times 1 are _____

Multiplication

I can use multiplication to work problems.

READ

Use objects to work the story problems in Lesson XXXIX. Remember that we are working with groups of **one** or **one group** of some number. Most of these problems are about money. We need to learn how to handle money wisely.

STUDY

John bought 2 figs at 1¢ each. How much did they cost?

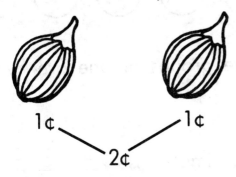

2 figs at 1¢ each.
2 times 1 are 2

RECITE

Use objects if necessary to answer all questions 1 through 9 in Lesson XXXIX, page 42. Ask your teacher/tutor to ask **you** the questions.

Record your answers here.

1. _____2 cents_____

2. _____   6. _____

3. _____   7. _____

4. _____   8. _____

5. _____   9. _____

Review VI

Addition and subtraction are combining or removing single objects.

6 and 8 are 14

14 less 6 leaves 8

When we multiply we are not using single objects, but groups of objects. The group may have only one object

in it. We may be talking about only one group. Notice that each one has the same answer.

one group of five

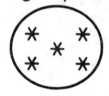

one group of five is five

five groups of one

five groups of one are five

Write the numbers on the lines.

1. 1 time 1 is _____

2. 1 time 2 is _____

3. 1 time 3 is _____

4. 1 time 4 is _____

5. 1 time 5 is _____

6. 1 time 6 is _____

7. 1 time 7 is _____

8. 1 time 8 is _____

9. 1 time 9 is _____

10. 1 time 10 are _____

11. 1 time 1 is _____

12. 2 times 1 are _____

13. 3 times 1 are _____

14. 4 times 1 are _____

15. 5 times 1 are _____

16. 6 times 1 are _____

17. 7 times 1 are _____

18. 8 times 1 are _____

19. 9 times 1 are _____

20. 10 times 1 are _____

Quiz VI

Write the numbers on the lines.

1. 1 time   5 is   _____

2. 6 times   1 are   _____

3. 1 time   10 is   _____

4. 4 times   1 are   _____

5. 8 times   1 are   _____

6. 1 time   6 is   _____

7. 9 times   1 are   _____

8. 2 times   1 are   _____

9. 10 times   1 are   _____

10. 1 time   7 is   _____

11. 3 nests each having 2 eggs are _____ eggs.

12. 2 boxes each having 2 pencils are _____ pencils.

13. Jan's hand holds 3 nickels which is _____ cents.

14. 2 strings of beads with 4 beads each are _____ beads.

15. In Dad's pocket is 10 pennies which is _____ cents.

Answer five oral questions from Lesson XXXIX. These questions should be asked by a teacher/tutor.

Multiplication

I can learn multiplication by two.

READ

When we do multiplication, we are working with groups. In Lesson XL, page 43, we are working with groups of **two**, or **two groups** of some number.

Read the sentences at the top of page 43. Notice that there are two sentences with the same answer. Practice these sentences using objects.

STUDY

2 times 3 are 6
2 groups of 3 are 6

3 times 2 are 6
3 groups of 2 are 6

RECITE

Write the numbers on the lines.

1. 2 times 1 are _____

2. 2 times 2 are _____

3. 2 times 3 are _____

4. 2 times 4 are _____

5. 2 times 5 are _____

6. 2 times 6 are _____

7. 2 times 7 are _____

8. 2 times 8 are _____

9. 2 times 9 are _____

10. 2 times 10 are _____

11. 1 time 2 is _____

12. 2 times 2 are _____

13. 3 times 2 are _____

14. 4 times 2 are _____

15. 5 times 2 are _____

16. 6 times 2 are _____

17. 7 times 2 are _____

18. 8 times 2 are _____

19. 9 times 2 are _____

20. 10 times 2 are _____

Multiplication

I can use multiplication to work problems.

READ

Continue to use objects to work the story problems in Lesson XL. Remember that we are working with groups of **two** or **two groups** of some number.

Most of the problems are about money. We must learn to handle money correctly and wisely.

STUDY

Peaches are selling at 2 cents each.
How much will 2 peaches cost?

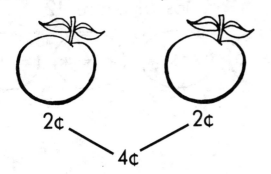

2 peaches at 2¢ each
2 times 2 are 4

RECITE

Use objects if necessary to answer all questions 1 through 9 in Lesson XL, page 43. Ask your teacher/tutor to ask **you** the questions.

Record your answers here.

1. _____ 4 cents _____

2. _____

6. _____

3. _____

7. _____

4. _____

8. _____

5. _____

9. _____

Multiplication

I can learn multiplication by three.

READ

Remember that when we do multiplication, we are working with groups. In Lesson XLI, page 44, we are working with groups of **three**, or **three groups** of some number. Read the sentences at the top of page 44. Notice that there are two sentences with the same answer. Practice these sentences using objects.

STUDY

3 times 4 are 12
3 groups of 4 are 12

4 times 3 are 12
4 groups of 3 are 12

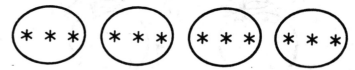

RECITE

Write the numbers on the lines.

1. 3 times 1 are _____

2. 3 times 2 are _____

3. 3 times 3 are _____

4. 3 times 4 are _____

5. 3 times 5 are _____

6. 3 times 6 are _____

7. 3 times 7 are _____

8. 3 times 8 are _____

9. 3 times 9 are _____

10. 3 times 10 are _____

11. 1 time 3 is _____

12. 2 times 3 are _____

13. 3 times 3 are _____

14. 4 times 3 are _____

15. 5 times 3 are _____

16. 6 times 3 are _____

17. 7 times 3 are _____

18. 8 times 3 are _____

19. 9 times 3 are _____

20. 10 times 3 are _____

Multiplication

I can use multiplication to work problems.

READ

Using objects, work the problems in Lesson XLI. Remember that we are working with groups of **three** or **three groups** of some number.

Most of these story problems are about money. We must learn to handle money correctly and wisely.

STUDY

A pint of chestnuts cost 3 cents. How much will 2 pints cost?

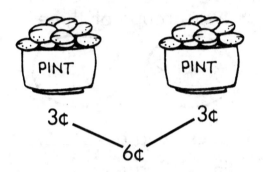

2 pints at 3¢ each.
2 times 3 are 6

RECITE

Use objects if necessary to answer all questions 1 through 9 in Lesson XLI, page 44. Ask your teacher/tutor to ask **you** the questions.

Record your answers here.

1. _____ 6 cents _____

2. _____   6. _____

3. _____   7. _____

4. _____   8. _____

5. _____   9. _____

Review VII

When we multiply, we are using groups of objects. The groups may have only **one** object in it or we may be talking about **one** group.

The group may have **two** objects in it. We may be talking about **two** groups of some number.

two groups of four

two groups of four are eight

four groups of two

four groups of two are eight

The group may have **three** objects in it. We may be talking about **three** groups of some number.

three groups of two

two groups of three

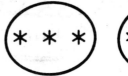

Write the numbers on the lines.

1. 2 times 2 are _____
2. 2 times 3 are _____
3. 2 times 4 are _____
4. 2 times 5 are _____
5. 2 times 6 are _____
6. 2 times 7 are _____
7. 2 times 8 are _____
8. 2 times 9 are _____
9. 2 times 10 are _____
10. 3 times 2 are _____
11. 3 times 3 are _____
12. 3 times 4 are _____
13. 3 times 5 are _____
14. 3 times 6 are _____
15. 3 times 7 are _____

16. 3 times 8 are _____
17. 3 times 9 are _____
18. 3 times 10 are _____
19. 4 times 2 are _____
20. 5 times 2 are _____
21. 6 times 2 are _____
22. 7 times 2 are _____
23. 8 times 2 are _____
24. 9 times 2 are _____
25. 4 times 3 are _____
26. 5 times 3 are _____
27. 6 times 3 are _____
28. 7 times 3 are _____
29. 8 times 3 are _____
30. 9 times 3 are _____

Quiz VII

Write the numbers on the lines.

1. 2 times 2 are _____   **11.** 4 times 3 are _____

2. 3 times 4 are _____   **12.** 9 times 2 are _____

3. 4 times 2 are _____   **13.** 7 times 3 are _____

4. 2 times 3 are _____   **14.** 3 times 8 are _____

5. 6 times 2 are _____   **15.** 2 times 10 are _____

6. 2 times 7 are _____   **16.** 2 times 9 are _____

7. 3 times 5 are _____   **17.** 3 times 7 are _____

8. 8 times 2 are _____   **18.** 2 times 8 are _____

9. 2 times 5 are _____   **19.** 3 times 9 are _____

10. 3 times 3 are _____   **20.** 3 times 10 are _____

21. 3 nests each having 5 eggs are _____ eggs.

22. 4 boxes each having 2 pencils are _____ pencils.

23. Ted's hand holds 2 dimes which is _____ cents.

24. 3 strings of beads with eight beads each are _____ beads.

25. In Mother's handbag is 2 nickels which is _____ cents.

Answer five oral questions from Lessons XL or XLI, pages 43 or 44. These questions should be asked by a teacher/tutor.

Multiplication

I can learn multiplication by four.

READ

When we do multiplication, we are working with groups. In Lesson XLII, page 45, we are working with groups of **four**, or **four groups** of some number.

Read the sentences at the top of page 45. Notice that there are two sentences with the same answer. Practice these sentences using objects.

STUDY

4 times 2 are 8
4 groups of 2 are 8

2 times 4 are 8
2 groups of 4 are 8

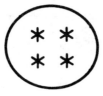

RECITE

Write the numbers on the lines.

1. 4 times 1 are _____

2. 4 times 2 are _____

3. 4 times 3 are _____

4. 4 times 4 are _____

5. 4 times 5 are _____

6. 4 times 6 are _____

7. 4 times 7 are _____

8. 4 times 8 are _____

9. 4 times 9 are _____

10. 4 times 10 are _____

11. 1 time 4 is _____

12. 2 times 4 are _____

13. 3 times 4 are _____

14. 4 times 4 are _____

15. 5 times 4 are _____

16. 6 times 4 are _____

17. 7 times 4 are _____

18. 8 times 4 are _____

19. 9 times 4 are _____

20. 10 times 4 are _____

Multiplication

I can use multiplication to work problems.

READ

Continue to use objects to work the story problems in Lesson XLII. Remember that we are working with groups of **four** or **four groups** of some number.

Most of the problems are about money. We must learn to handle money correctly and wisely.

STUDY

Lucy has 2 kittens.
Each kitten has 4 feet.
How many feet have both kittens?

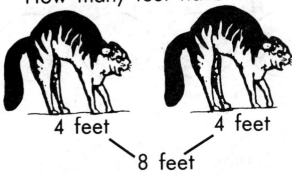

4 feet   4 feet
        8 feet

2 kittens with 4 feet each.
2 times 4 are 8

RECITE

Use objects if necessary to answer all questions 1 through 9 in Lesson XLII, page 45. Ask your teacher/tutor to ask **you** the questions.

Record your answers here.

1. _____8 feet_____

2. _____   6. _____

3. _____   7. _____

4. _____   8. _____

5. _____   9. _____

Multiplication

I can learn multiplication by five.

READ

Remember that when we do multiplication, we are working with groups. In Lesson XLIII, page 46, we are working with groups of **five** or **five groups** of some number. Read the sentences at the top of page 46. Notice that there are two sentences with the same answer. Practice these sentences using objects.

STUDY

5 times 4 are 20
5 groups of 4 are 20

4 times 5 are 20
4 groups of 5 are 20

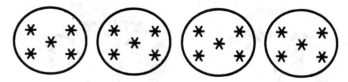

RECITE

Write the numbers on the lines.

1. 5 times 1 are _____
2. 5 times 2 are _____
3. 5 times 3 are _____
4. 5 times 4 are _____
5. 5 times 5 are _____
6. 5 times 6 are _____
7. 5 times 7 are _____
8. 5 times 8 are _____
9. 5 times 9 are _____
10. 5 times 10 are _____

11. 1 time 5 is _____
12. 2 times 5 are _____
13. 3 times 5 are _____
14. 4 times 5 are _____
15. 5 times 5 are _____
16. 6 times 5 are _____
17. 7 times 5 are _____
18. 8 times 5 are _____
19. 9 times 5 are _____
20. 10 times 5 are _____

Multiplication

I can use multiplication to work problems.

READ

Using objects, work the problems in Lesson XLIII. Remember that we are working with groups of **five** or **five groups** of some number.

Most of these story problems are about money. We must learn to handle money correctly and wisely.

STUDY

Francis bought 5 tops at 2 cents each. How many cents do they cost?

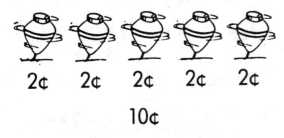

2¢   2¢   2¢   2¢   2¢

10¢

5 tops at 2¢ each.
5 times 2 are 10

RECITE

Use objects if necessary to answer all questions 1 through 9 in Lesson XLIII, page 46. Ask your teacher/tutor to ask **you** the questions.

Record your answers here.

1. _____10 cents_____

2. _____

3. _____

4. _____

5. _____

6. _____

7. _____

8. _____

9. _____

Review VIII

When we multiply, we are using groups of objects. The group may have only one, two, or three objects in it. We may be talking about one, two, or three groups of some number.

The group may have four objects in it. We may be talking about four groups of some number.

four groups of three

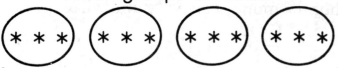

three groups of four

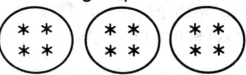

The group may have five objects in it. We may be talking about five groups of some number.

five groups of two

two groups of five

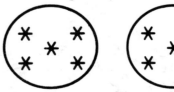

Write the numbers on the lines.

1. 4 times 2 are _____
2. 4 times 3 are _____
3. 4 times 4 are _____
4. 4 times 5 are _____
5. 4 times 6 are _____
6. 4 times 7 are _____
7. 4 times 8 are _____
8. 4 times 9 are _____
9. 4 times 10 are _____
10. 5 times 2 are _____
11. 5 times 3 are _____
12. 5 times 4 are _____
13. 5 times 5 are _____
14. 5 times 6 are _____
15. 5 times 7 are _____
16. 5 times 8 are _____
17. 5 times 9 are _____
18. 5 times 10 are _____
19. 4 times 4 are _____
20. 5 times 4 are _____
21. 6 times 4 are _____
22. 7 times 4 are _____
23. 8 times 4 are _____
24. 9 times 4 are _____
25. 4 times 5 are _____
26. 5 times 5 are _____
27. 6 times 5 are _____
28. 7 times 5 are _____
29. 8 times 5 are _____
30. 9 times 5 are _____

Quiz VIII

Write the numbers on the lines.

1. 4 times 2 are _____

2. 5 times 4 are _____

3. 4 times 4 are _____

4. 4 times 3 are _____

5. 6 times 4 are _____

6. 4 times 7 are _____

7. 5 times 5 are _____

8. 8 times 4 are _____

9. 4 times 5 are _____

10. 5 times 3 are _____

11. 3 times 5 are _____

12. 9 times 4 are _____

13. 7 times 5 are _____

14. 5 times 8 are _____

15. 4 times 10 are _____

16. 4 times 9 are _____

17. 5 times 7 are _____

18. 4 times 8 are _____

19. 5 times 9 are _____

20. 5 times 10 are _____

21. 5 nests each having 4 eggs are _____ eggs.

22. 6 boxes each having 5 pencils are _____ pencils.

23. Joy holds a dime in each hand which is _____ cents.

24. 4 strings of beads with seven beads each are _____ beads.

25. Father had 5 nickels in his pocket which is _____ cents.

Answer five oral questions from Lessons XLII or XLIII, pages 45 or 46. These questions should be asked by a teacher/tutor.

Review I-II

Subtraction is removing objects from the first number. The words **from** and **leaves** are used in subtraction.

When we subtract one from the first number, we have one less for the answer.

When we subtract two from the first number, we have two less for the answer.

When we subtract three from the first number, we have three less for the answer.

When we subtract four from the first number, we have four less for the answer.

When we subtract five from the first number, we have five less for the answer.

When we subtract six from the first number, we have six less for the answer.

When we subtract seven from the first number, we have seven less for the answer.

1 from 5 leaves 4   4 from 6 leaves 2

2 from 4 leaves 2   5 from 9 leaves 4

3 from 7 leaves 4   6 from 7 leaves 1

7 from 15 leaves 8

Quiz I-II

Write the numbers on the lines.

1. 3 from 10 leaves _____
2. 4 from 6 leaves _____
3. 6 from 14 leaves _____
4. 1 from 8 leaves _____
5. 3 from 4 leaves _____
6. 2 from 9 leaves _____
7. 6 from 8 leaves _____
8. 3 from 9 leaves _____
9. 2 from 10 leaves _____
10. 7 from 15 leaves _____

11. 6 from 9 leaves _____
12. 5 from 9 leaves _____
13. 3 from 11 leaves _____
14. 6 from 15 leaves _____
15. 4 from 12 leaves _____
16. 5 from 14 leaves _____
17. 7 from 13 leaves _____
18. 6 from 11 leaves _____
19. 3 from 10 leaves _____
20. 5 from 12 leaves _____

21. Five of eleven birds flew away. _____ birds were left.

22. Eight oranges taken from fourteen oranges leaves _____ oranges.

23. John spent eight cents of fifteen cents leaving _____ cents.

24. Wendy lost seven of sixteen marbles leaving _____ marbles.

25. Nine grapes from seventeen grapes leaves _____ grapes.

Answer 10 oral questions from Lessons XXV, through XXXI. These questions should be asked by a teacher/tutor.

51 (fifty-one)

Review III-IV

Subtraction is removing objects from the first number. The words **from** and **leaves** are used in subtraction. We ask the question how many are left.

When we subtract eight from the first number, we have eight less for the answer.

When we subtract nine from the first number, we have nine less for the answer.

When we subtract ten from the first number, we have ten less for the answer.

We learn to work story problems using subtraction. Story problems are written in English words. We must read these problems carefully, set the problem up, and then find the answer.

Besides the use of the word **from**, we also use the word **less** in subtraction.

Some problems use both addition and subtraction. The word **and** tells us to add and the word **less** tells us to subtract. When we add, we are counting forward. When we subtract, we are counting in reverse, backward.

8 from 12 leaves 4   12 less 8 are 4

9 from 16 leaves 7   16 less 9 are 7

10 from 14 leaves 4   14 less 10 are 4

3 and 6 less 4 are 5

5 and 7 less 3 are 9

Quiz III-IV

Write the numbers on the lines.

1. 8 from 13 leaves _____
2. 9 from 16 leaves _____
3. 5 from 15 leaves _____
4. 8 from 17 leaves _____
5. 6 from 15 leaves _____
6. 9 from 13 leaves _____
7. 5 from 14 leaves _____
8. 10 from 17 leaves _____
9. 9 from 18 leaves _____
10. 8 from 12 leaves _____

11. 17 less 7 are _____
12. 14 less 6 are _____
13. 9 less 3 are _____
14. 13 less 7 are _____
15. 11 less 5 are _____
16. 17 less 9 are _____
17. 11 less 7 are _____
18. 13 less 6 are _____
19. 14 less 8 are _____
20. 16 less 9 are _____

21. 7 and 9 less 10 are _____
22. 6 and 7 less 3 are _____
23. 2 and 7 less 4 are _____
24. 5 and 5 less 2 are _____
25. 4 and 9 less 5 are _____

Answer ten oral questions from Lesson XXXII through XXXIV. These questions should be asked by a teacher/tutor.

Review V-VI

When we add, we use the word **and**. We count forward to find the answer in addition. When we subtract, we use the word **less**. We count in reverse, or backward, to find the answer in subtraction.

We can work addition and subtraction in the same problem. We add when we see the word **and** and we subtract when we see the word **less**.

Problems can be written in English words. When problems are written in words, we must read carefully. Then we write the problem in numbers. Finally we find the answer.

7 and 6 less 4 are 9

8 and 9 less 7 are 10

We are also learning to work with groups of numbers. Each group has the same number. We can have one group of six or we can have six groups of one object each. Each problem has the same answer.

One group of six

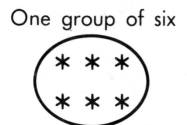

Six groups of one

Quiz V-VI

Write the numbers on the lines.

1. 4 and 8 less 5 are _____
2. 5 and 6 less 4 are _____
3. 1 and 9 less 7 are _____
4. 10 and 6 less 9 are _____
5. 6 and 9 less 7 are _____

6. 2 and 8 less 5 are _____
7. 5 and 8 less 6 are _____
8. 10 and 6 less 8 are _____
9. 9 and 5 less 7 are _____
10. 10 and 3 less 8 are _____

11. 1 time 4 is   _____
12. 3 times 1 are   _____
13. 2 times 1 are   _____
14. 1 time 6 is   _____
15. 10 times 1 are   _____

16. 1 time   8 is   _____
17. 9 times   1 are   _____
18. 5 times   1 are   _____
19. 1 time 10 is   _____
20. 6 times   1 are   _____

21. 4 nests each having 3 eggs are _____ eggs.

22. One box with 6 pencils is _____ pencils.

23. In Dad's pocket is 1 dime which is _____ cents.

24. 2 strings of beads with 3 beads each are _____ beads.

25. I think of two numbers that together make 15. One of the numbers

    is 6. What is the other number? _____

Answer five oral questions from Lesson XXXIX. These questions should be asked by a teacher/tutor.

55 (fifty-five)

Review VII-VIII

When we multiply, we are using groups of objects. Each group has the same number of objects. The group may have two objects or two groups may have some number of objects.

One group may have **three** objects in it. We may be talking about **three** groups of some number of objects.

One group may have **four** objects in it. We may be talking about **four** groups of some number of objects.

One group may have **five** objects in it. We may be talking about **five** groups of some number of objects.

2 groups of 3

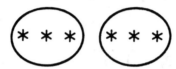

3 groups of 2

5 groups of 4

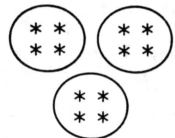

4 groups of 5

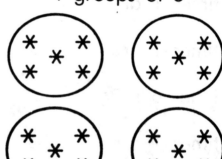

3 groups of 4

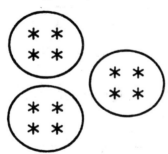

4 groups of 3

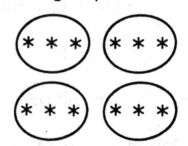

Quiz VII-VIII

Write the numbers on the lines.

1. 4 times 2 are _____   **11.** 4 times 9 are _____

2. 7 times 3 are _____   **12.** 7 times 4 are _____

3. 5 times 2 are _____   **13.** 5 times 5 are _____

4. 6 times 3 are _____   **14.** 4 times 4 are _____

5. 8 times 4 are _____   **15.** 4 times 6 are _____

6. 5 times 8 are _____   **16.** 4 times 3 are _____

7. 3 times 9 are _____   **17.** 3 times 3 are _____

8. 2 times 6 are _____   **18.** 5 times 6 are _____

9. 5 times 7 are _____   **19.** 2 times 2 are _____

10. 4 times 5 are _____   **20.** 3 times 8 are _____

21. 5 nests each having 3 eggs are _____ eggs.

22. 4 boxes each having 8 pencils are _____ pencils.

23. Kara has six nickels. She has _____ cents.

24. 3 strings of beads with nine beads on a string are _____ beads.

25. Father has nine nickels. Father has _____ cents.

Answer ten oral questions from Lessons XL or XLIII, pages 43 to 46.
These questions should be asked by a teacher/tutor.

57 (fifty-seven)

NOTES

NOTES

NOTES

NOTES

NOTES

NOTES

NOTES

NOTES

NOTES

NOTES